AF887999

Who Am I?

Jungle Animals

Written by Read With You
Center for Language Research and Development

The characters and events portrayed in this book are fictitious. Any similarity to real persons, living or dead, is coincidental and not intended by the author.

Published by Read With You Publishing

Designed by Read With You Center for Language Research and Development

ISBN-13: 978-1-944710-79-8
ISBN-10: 1-944710-79-5

Printed in the United States of America.

I have a row of spines running down my back and tail.

Who am I?

iguana

I use rocks to crack open nuts.

Who am I?

chimpanzee

I can grasp things with both my fingers and toes.

Who am I?

monkey

I have a super long tongue.

Who am I?

anteater

My arms are longer than
my legs.

Who am I?

gorilla

I sleep upside down.

Who am I?

bat

Instead of laying eggs, I give birth to live babies.

Who am I?

boa constrictor

I can regrow lost legs.

Who am I?

tarantula

Iguana

Chimpanzee

Monkey

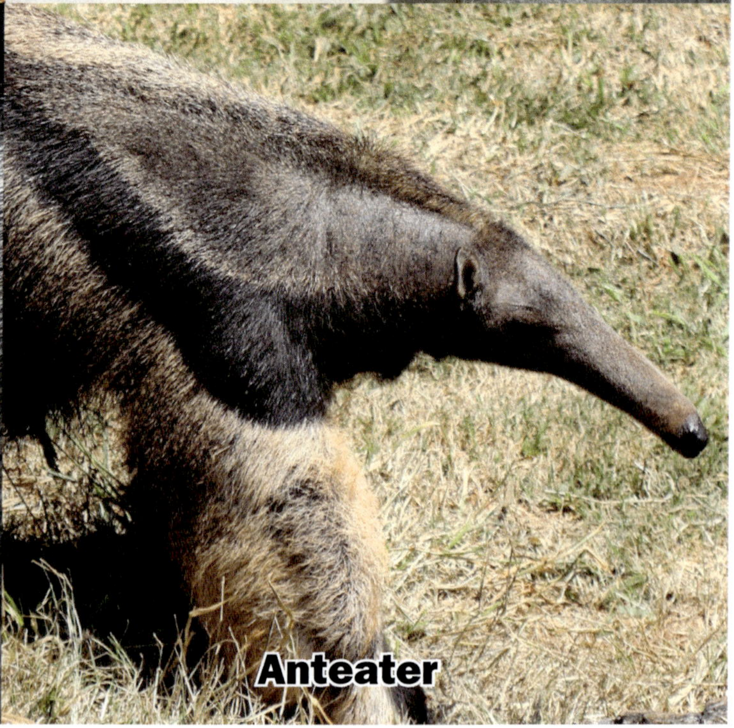
Anteater

I am an iguana.

I am a chimpanzee.

I am a monkey.

I am an anteater.

Gorilla

Bat

Boa Constrictor

Tarantula

I am a gorilla.

I am a bat.

I am a boa constrictor.

I am a tarantula.

We are Jungle animals.

A jungle is a forest thick with trees, plants, and animals.

Made in United States
Orlando, FL
07 November 2026

90337520R00017